SCALING AI

THE AI GOVERNANCE AND SECURITY
PLAYBOOK FOR EXECUTIVES

SOL RASHIDI,
Chief Strategy Officer for Data & AI

STEVE KLEMENTOWSKI,
Principal Solution Architect

Published by Atlas Elite Publishing Partners
Interior design by Michael Beas

Paperback ISBN: 978-1-972014-01-1

Printed in United States of America

For more information, visit:
www.atlaselitepublishingpartners.com

TABLE OF CONTENTS

FORWARD

We are standing at a defining moment in technology leadership. Artificial Intelligence (AI) is no longer an experimental capability sitting in research or at the edge of business—it's here! And with this power comes profound responsibility, as AI is rapidly becoming the engine that drives competitiveness, customer experience, and growth, embedding itself into all core processes. With this opportunity also comes risk. AI expands our attack surface, challenges our governance models, and tests the limits of our regulatory frameworks. Leaders can no longer afford to treat security, compliance, and ethics as parallel tracks to innovation; they must be embedded from the very start, and we must ensure that the systems we design, deploy, and depend upon are secure, compliant, and worthy of the trust placed in them.

That's why we wrote this eBook, a condensed and concise playbook written for the leaders at the helm of that responsibility—leaders such as the **Chief Information Security Officers, Chief Data Officers, Chief Information**

Officers, Chief Technology Officers, and Chief AI Officers—who are now the architects of safe and scalable AI. Whether you are an established executive guiding a global AI program or an emerging leader preparing to take on the mantle, the mission is the same: to shape the future of AI in a way that is secure and sustainable.

Drawing on proven frameworks, regulatory best practices, and defensible strategies, this eBook provides you with actionable roadmaps, frameworks, and models for governing AI across its full lifecycle—from data acquisition and classification, to model deployment, monitoring, and continuous improvement. It equips you to navigate the complexity of:

- **Securing sensitive data** against leakage, misuse, and unauthorized access.
- **Embedding governance and compliance** into AI pipelines from day one.
- **Protecting foundation models** from adversarial manipulation and drift.
- **Measuring and reducing risk** without slowing innovation.
- **Preparing the workforce** to operate AI responsibly and effectively.

This eBook is not filled with a bunch of theory! It's from experience. It is a field manual for leaders who understand that their role is not just to adopt AI, but to shape how AI is built, secured, and governed for years to come; it is an

2

enterprise-wide imperative, rooted in resilience and foresight.

The truth is, decisions you make in the next 12 to 24 months will echo for a decade. They will influence not only your organization's competitiveness but also its resilience, its reputation, and its trustworthiness in the eyes of customers, partners, and regulators. Security is not the cost of doing business with AI; it is the currency that earns trust from customers, regulators, and shareholders alike, and you, dear leaders, are at the helm!

This eBook is your blueprint for leading AI into its next chapter, and we encourage you to lead with speed, but never without safeguards. Innovate boldly, but govern relentlessly. Because in the age of AI, **what you do matters—and how you do it matters even more.**

— Sol Rashidi, Cyera, Chief Strategy Officer for Data & AI

EXECUTIVE SUMMARY

We're entering an era where data isn't just the new oil —it serves as the underlying engine to AI's transformation of nearly every business process in a variety of industries and sectors. As organizations race to GenAI, Applied AI, or Agentic AI across functions, the new C-Suite such as the Chief Data Officer (CDO), Chief Information Security Officer (CISO), Chief Information Officer (CIO), Chief Artificial Intelligence Officer (CAIO), Chief Technology Officer (CTO), and many more have shifted from back-office enablement functions to front-line innovation drivers. This is the moment where data strategy, business strategy, and security become inseparable. From Business Intelligence (BI) to Artificial Intelligence (AI) and all operational aspects in between, this is the new power grid of the 21st century.

This eBook explores the practical and urgent imperatives for the modern-day C-Suite, technical executives, or any

leader who's been trusted with scaling AI in a safe, secure, and governed way. It offers an unfiltered look at the challenges, the opportunities, and how the new modern heroes of today— our security and governance specialists— can turn intent into impact.

This eBook outlines nine critical considerations, structured into chapters, for the modern-day CXO, in safely unlocking value from AI while securing it for scalability. We will discuss the honest challenges that come with it, real use-cases, and how a proactive data-centric approach empowers organizations to innovate responsibly and competitively.

CHAPTER 1

The New AI Mandate for Modern-Day CXOs

> *According to McKinsey, AI could add up to $4.4 trillion annually to the global economy[1]. Gartner reports that more than 75% of enterprises plan to integrate AI into at least one function within the next 12 months due to this phenomenon[2].*

In the age of AI, it must be understood that **Data Discovery** (which unlocks opportunities to innovate, personalize, and scale), **Data Classification** (which unlocks opportunities to observe, identify, and source

---

[1] McKinsey Global Institute. *The Economic Potential of Generative AI: The Next Productivity Frontier.* McKinsey & Company, June 14, 2023. https://www.mckinsey.com/.
[2] CIO Dive. "Gartner: Three-Quarters of Enterprises Will 'Operationalize' AI by 2024." June 22, 2020. https://www.ciodive.com/.

usage and access), and **Data Governance** (which unlocks opportunities to protect, empower, and future-proof your strategy) are the three facets to scaling AI healthily and sustainably. Without it, it would be like driving a Ferrari on the speedway without a seatbelt and safety gear—no one would dare, including race car drivers.

And when you combine these three facets with a sound AI strategy, these become the pillars that form the blueprint for your **Data Intelligence**. This ecosystem doesn't just account for storage, but also for strategy, security, and scale, which ultimately enable you to respond to business needs and capabilities in a fast and secure manner. As we move forward, our mental models need to shift from saying 'NO' or 'YES', to 'YES, and here's HOW' to stay valuable to the business. In today's AI-powered world, algorithms may drive the car, but data decides the destination. AI might be the prom king, dazzling with moves and flair, but data? Data is still the prom queen— quietly commanding the room, setting the tone, and stealing the crown.

That now leaves us with this well-known fact: data is, and will always be, an operational imperative that's shifted from an unglamorous back-end function in the early 2000s, to now taking center stage.

For Chief Data Officers (CDOs), Chief Information Security Officers (CISOs), Chief Artificial Intelligence Officers (CAIOs), Chief Information Officers (CIOs), Chief

Technology Officers (CTOs), and more, this is a defining moment. We are now at the nexus of AI strategy and data enablement, uniquely positioned to lead organizations through both disruption and transformation. However, this immense potential also comes with significant risks that organizations must address head-on.

As the adoption of AI accelerates, leaders must navigate a complex landscape of challenges, including the risks of hallucination and bias, as well as data leaks and prompt poisoning through accidental and unintentional data exposure via prompts from either AI, shadow AI, or unsanctioned AI use.

Furthermore, foundational models like GPT and Mistral are only as powerful as the data that fuels them since these models rely on vast quantities of high-quality, well-governed, and compliant data to drive insights, generate content, and automate decisions. That's a concept we all understand and adhere to. In reality, however, our data is usually lacking in quality and accessibility, making our roles within enterprises more challenging as we try to manage the hallucinations, data leaks, misuse, and prompt injection attacks.

Nonetheless, that hasn't stopped the evolution of these models or their use!

> **"This marks a true inflection point—a new era. The C-Suite is shifting from acting as the organization's data custodians to becoming the architects of its future and enterprise intelligence."**

The C-Suite sits at the intersection of technology, regulation, and business value. And while there are many frameworks, models, and tools to choose from, what's foundational in all decisions and choices moving forward is operationalizing visibility, classification, and control at scale since we are now expected to:

1. Operationalize trustworthy data pipelines to power AI models.
2. Classify and protect sensitive data to avoid leakage into prompts or outputs.
3. Ensure compliance with rapidly evolving data privacy and AI governance regulations.
4. Democratize access to high-value data sets for AI innovation.
5. Define value metrics that link AI-driven outcomes back to data investments.

CDOs, CISOs, CIOs, CAIOs, and CTOs combined must architect a data foundation that balances innovation with control and safety while unlocking business value. It's truly an exercise in art and science, and there is no one-size-fits-all solution. Factors such as industry, sector, and corporate culture all play a role in understanding where that balance lies. So, the goal of this eBook is to ensure our executives who are responsible for architecting this forward-looking view of AI for their organizations are empowered with sound mental models to ensure they don't stray from the three operational imperatives provided below:

Pillar	Description
Classification	Manual classification is no longer an option; data must be auto-classified with AI-native capabilities that highlight the context of the data, the users, and business associations quickly and accurately, while taking into consideration compliance for governance, protection, and downstream AI use. Without this, you can't get to the 2nd operational imperative, Observability.
Observability	With data classification in place, you can now establish end-to-end visibility into data flows, model usage, API calls, and AI tool integrations across the stack. When you have observability, you can then begin to Control, the third operational imperative.
Control	To establish control, you should implement guardrails at data ingestion, model API endpoints, prompt/response layers, and monitor usage in real-time. Long gone are the days of manual permission-based access and control.

CHAPTER 2

Frameworks, Metrics, and Models that Guide You

Within this eBook, we've outlined frameworks, models, and maturity models. All of which are critical for your AI initiatives, because as your enterprises scale AI across functions and embed it into core workflows, your attack surface expands—from data sourcing to model outputs. Without a prepared team and a structured end-to-end approach to securing each phase of the AI lifecycle, organizations risk data leaks, regulatory non-compliance, model manipulation, and reputational damage.

As such, we've worked on providing frameworks and models that offer a pragmatic, phased methodology for systematically hardening every layer of AI deployment, ensuring resilience, compliance, and trust while preserving the flexibility needed to innovate at enterprise speed.

To start, we'll share with you the *7-Phase AI Security Framework* and *AI-Security ROI Metric System*, created by Sol Rashidi from Cyera, the leading AI and data security

company. Outlined below are the different pillars needed to secure your AI ecosystem end-to-end. The seven phases include:

Phase	Focus	Key Controls
1. Data Sourcing	Discovery, classification, provenance tracking, legal rights, and vendor risk	Data registry, inventory, compliance checker
2. Infrastructure	Segment networks, encrypt flows, and container security	Terraform configs, KMS, TLS, VPC isolation
3. Data In-Transit	Masking, anonymization, and differential privacy	PII detection, synthetic data checks
4. API	Authenticate access to models and enforce rate limits	RBAC, token validation, input sanitization
5. Model Provenance	Detect prompt injection, adversarial behavior, and unsanctioned use	Output filters, RAG, guardrails
6. Incident Response	Real-time alerts, forensic lineage, and rollback plans.	Automated breach response, alert triage
7. Continuous Monitoring	Drift detection, usage audits, compliance logging	KPI dashboards, retraining triggers

In addition to the *7-Phase AI Security Framework*, which provides comprehensive protection of your AI capabilities, there is also an *AI-Security ROI Metric System*, which is essential in aligning the gap between AI investments and measurable business outcomes. While the *7-Phase AI*

Security Framework will ensure you've thought through your AI lifecycle end-to-end, your teams will still need to quantify the ROI of AI initiatives and their spend on security initiatives so that security and data do not become sunk costs and checkbox exercises.

The **AI-Security ROI Metric System,** created for CDOs, CISOs, CTOs, CIO s, and CAIOs, offers executives an objective and quantifiable way to transform AI from a technical experiment into a strategic asset by linking specific datasets, models, and use cases to tangible results, like revenue lift, cost savings, or risk reduction. It empowers CDOs, CISOs, and CAIOs to justify data and security spending, optimize future investments, and clearly communicate value to boards and executive teams.

Metrics to leverage include:

Metric	Description	Example
Data-to-Model Yield	% of usable data converted into model-ready format	Being able to say "17% of enterprise data is model-ready"
Prompt Leakage Rate	% of sensitive data detected in prompt streams	Real-time alerting through Cyera's runtime AI API
Model ROI Attribution	Business outcomes directly tied to specific datasets	Linking sales uplift to personalization model inputs
Prompt Leakage Rate	% of AI model prompts or outputs containing sensitive data	Reports that evaluate runtime data loss risk
Model Access Governance Score	% of API endpoints governed by RBAC, rate limits, and audit logging	Measures runtime control implementation
Lineage Traceability Score	% of AI outputs that can be traced to their original data source	Supports forensic audits and regulatory compliance
Time-to-Trust (T2T)	Days from model completion to achieving trust/compliance for deployment	Measures security & governance readiness
Model Deployment Velocity	Number of production-ready models shipped per quarter	Indicates operational efficiency of AI teams
Retraining Frequency	Number of model refresh cycles per year	Connects data drift and model staleness
Cost-to-Serve Reduction via AI	Savings from AI automation in ops, support, or decision-making	Tracks ROI from applied AI use cases
Data Investment Efficiency	Ratio of AI-driven outcomes to total data infrastructure spend	Tracks ROI for data spend

The ***AI-Security ROI Metric System*** outlined above
provides a starting point for CDOs, CISOs, CIOs, and
CAIOs to quantify the interplay between data assets, AI
models, and security posture. Still, it is by no means a
comprehensive list or a rigid blueprint. These metrics—
ranging from Data-to-Model Yield to Data Investment
Efficiency —can be adopted wholesale for organizations
seeking immediate, standardized benchmarks. It's best,
however, to adapt them to reflect your unique operational
realities. For instance, a financial services firm might
enhance the Prompt Leakage Rate by integrating sector-
specific compliance thresholds under regulations like
GDPR or CCPA, while a healthcare provider could expand
Model ROI Attribution to include patient outcome
improvements tied to HIPAA-compliant datasets. The key
is flexibility: treat these as foundational building blocks
that evolve with your enterprise's maturity in AI adoption;
have full reign to adopt or adapt any of these metrics into
your KPIs and OKRs.

At its core, the **AI-Security ROI Metric System**, alongside
complementary frameworks such as AI governance
models or risk assessment protocols, serves as a compass
in the often tumultuous landscape of AI data security that
we are navigating together. In an era where AI is no longer
a siloed IT initiative but a boardroom imperative, these
guardrails and qualitative metrics prevent executives from
veering off course amid rapid technological shifts,
regulatory pressures, and emerging threats such as

adversarial attacks or data poisoning. Not to mention, having these measures in place builds immense credibility.

> **"The truth is, you can't manage what you can't measure, and the AI landscape has been plagued with subjective promises."**

Most importantly, you want to avoid any unwanted narrative that someone in your organization decides they want to spin because what you present is not in their favor. We've all been there! By using these metrics at the starting gate, you can offer clarity in chaos, enabling you to map out a direction that aligns AI investments with measurable value. Note: this will take a few iterations to figure out because the true north isn't universal; it has to be calibrated to account for your organization's internal compass points. So, adaptation of these metrics is not just encouraged—it's essential.

A good starting point is always to consider your company's culture. A startup with an agile, innovation-driven ethos might prioritize Model Deployment Velocity and Retraining Frequency to fuel rapid iteration. In contrast, a legacy enterprise in a heavily regulated industry could amplify Lineage Traceability Score and Time-to-Trust to emphasize auditability and compliance. Sector nuances play a pivotal role, too— manufacturing leaders might link AI Attributable Revenue Growth to predictive maintenance models, reducing downtime. At the same time, retail

executives adapt Cost-to-Serve Reduction to focus on supply chain efficiencies powered by AI forecasting. Industry-specific risks, such as intellectual property leaks in tech or patient privacy breaches in life sciences, demand that you tailor metrics like Model Access Governance Score to incorporate bespoke controls, perhaps integrating zero-trust architectures or advanced encryption standards.

> **"Remember, the goal isn't perfection out of the gate, but iterative progress, and this eBook will give you a running start. As I always say, Think BIG, start SMALL, then scale QUICKLY."**

Pilot a few metrics in a high-impact use case, gather feedback, and refine. Engage stakeholders early to ensure buy-in, and leverage tools like dashboards or automated reporting to make these KPIs living, breathing elements of your OKRs. In doing so, you'll not only justify AI expenditures but also build trust with boards, regulators, and customers, ensuring your organization navigates the AI frontier with confidence and foresight. Ultimately, these frameworks are your guideposts, illuminating the path forward while leaving room for the detours that define innovation in a dynamic world. As AI evolves, so too should your metrics— embrace them as allies in securing tomorrow's data-driven success.

CHAPTER 3

Common AI Adoption Paths: Navigating Risks and Controls

Artificial intelligence has already moved beyond the innovation labs and data science teams, embedding itself within enterprise. As outlined in Chapter 1, the new mandate for CXOs requires shifting from reactive governance to proactive enablement, ensuring that classification, observability, and control are operationalized at scale (of course this is easier said than done).

Regardless, the reality for executives is clear: AI adoption is already underway within organizations, sometimes sanctioned, often not.

The tricky thing is, there is no one path to AI adoption, it doesn't follow a single path; most commonly, it arrives in three ways:

- **Public tools** accessible to any employee with a browser.
- **Embedded features** within trusted SaaS applications.
- **Homegrown or blended solutions** built on foundational models.

For leaders, recognizing these entry points is the first step toward balancing innovation with control. Each pathway offers value, but each also introduces unique risks that demand visibility, governance, and accountability from the start. Let's visit each risk.

AI Adoption Path #1: External Public Tools

Publicly available GenAI tools such as ChatGPT, Perplexity, or Gemini represent the enterprise's fastest entry point for artificial intelligence. Their appeal is obvious: instant access to state-of-the-art foundational models without needing infrastructure, procurement cycles, or technical expertise. From a business perspective, this path offers immediate productivity gains. Employees use public tools to brainstorm ideas, draft email communications, summarize documents, or research new topics. For executives, this adoption path is often one of the first signs that AI is no longer an abstract future-state, but a present-day reality, actively reshaping how their teams work.

While the appeal with public tools is its convenience, it carries significant risks when viewed through a corporate lens. Sensitive data can be exposed if employees copy and paste internal content into prompts. Browser-based tools may bypass enterprise logging and audit functions, this essentially means leadership has little visibility into how, or how often, they are being used.

For executives, the challenge isn't deciding if public AI will be used inside the enterprise; it already is, but without visibility or controls. The real decision is whether leadership will set the guardrails now, defining what "responsible use" looks like, or allow adoption to outpace control. Addressing this issue requires not only security controls but also clarity about where these tools fit into the enterprise.

AI Adoption Path #2: AI Embedded in SaaS Applications

A second, and increasingly common adoption path is through embedded AI capabilities, AI features directly within existing business applications. Whether branded as Microsoft Copilot, Salesforce Einstein, or GitHub Copilot, these features appear inside platforms that employees already use daily. Often, they arrive through subtle product updates, switched on by default or masquerading as a beta feature. To the business user, they feel like natural extensions of existing workflows, unlocking efficiency that feels almost superhuman.

This embedded model can deliver compelling benefits. Existing tools become more powerful without requiring additional contracts, new vendors, or separate training. Sales teams can receive AI-driven recommendations in Salesforce, legal counsel can accelerate drafting and summarization inside Microsoft Office, and engineering teams can write code at break neck speed in AI enabled IDEs.

The adoption of these features introduces new forms of risk. Sensitive corporate data may flow through AI services without explicit approval or oversight, and since the model is controlled by the vendor, organizations must trust the provider's controls, model provenance, and security assurances. From a leadership perspective, the challenge with embedded AI is subtler than with public tools. Unlike publicly available consumer tools, embedded AI lives within the platforms you've already licensed and your employees rely on daily. This makes outright prohibition of embedded AI impractical and often counterproductive. Clear governance over contracts, vendor data processing, and user expectations becomes essential to balancing efficiency gains with accountability.

AI Adoption Path #3: Blended or Homegrown AI

The third adoption path is a blended or homegrown approach, where enterprises build their own AI capabilities using foundation models delivered through platforms such as AWS Bedrock, Azure OpenAI, or Google Vertex AI. In this adoption path, organizations integrate pre-trained models into custom workflows or chatbots tailored for their specific needs. This path blends the flexibility of public AI with the perceived safety of enterprise control, creating an opportunity to turn proprietary fine-tuning and context into a true competitive advantage.

From a business perspective, homegrown AI can be strategically powerful. It allows enterprises to tailor use cases to their unique business processes and risk appetite. Just as importantly, it allows organizations to harness their vast internal data sets, turning proprietary information into a differentiator rather than handing it over to external tools. The risks, however, are significant. These deployments often rely on complex APIs and platform services that require careful configuration. Missteps can expose sensitive data or create over-permissive access into your crown jewels. Because third parties develop many foundation models, enterprises inherit risks related to model provenance, hidden training data, or unexpected behaviors like hallucinations.

For executives, the challenge with blended AI is balancing opportunity with accountability and success requires coordination across IT, security, data, and compliance teams. Those organizations that approach this path deliberately can establish a foundation for responsible innovation; those that move ahead without guardrails risk creating fragile, unsustainable AI deployments.

Executive Considerations Across Adoption Paths

In looking forward, leaders should take the following actions to guide AI adoption responsibly across the enterprise.

- **Map AI entry points:** Identify where public tools, SaaS features, and homegrown solutions are being used across the org.
- **Set guardrails for public AI:** Publish clear "responsible use" policies, restrict sensitive data sharing, and provide approved alternatives.
- **Govern embedded AI in SaaS:** Review vendor contracts, require transparency on data use, and align user expectations with enterprise risk.
- **Control homegrown AI projects:** Mandate cross-functional oversight (IT, security, compliance, data science) and enforce configuration, access, and provenance checks.
- **Anchor in governance:** Build early foundations of data discovery, classification, and oversight to prevent shadow AI growth.

AI adoption is not a light switch you flip on or off; it's a spectrum. Enterprises will inevitably weave together a mix of external tools, embedded SaaS features, and homegrown deployments. Each adoption path brings undeniable advantages, but also hidden complexities.

CYERA

For today's C-suite, the mandate is clear: chart the course, and embed data security and governance from the outset. Left unchecked, AI will scale on its own, spreading like wildfire across teams and functions. But scaling securely, responsibly, and in alignment with long-term business strategy demands intentional leadership.

As AI spreads across your organization, it amplifies the very problems that already strain enterprise data: shadow data, poor visibility, inconsistent governance, and unstructured information at a massive scale. Chapter 4 explores these realities head-on, moving from reactive firefighting to building proactive measures that empower innovation without sacrificing trust.

CHAPTER 4

Navigating Data Security Challenges in the AI Era: From Reactive Friction to Responsive Resilience

> *Over 90% of enterprise data is unstructured, with much of it remaining unclassified or poorly documented, according to recent IDC estimates for 2025.[3]*

This staggering volume —projected to reach 175 zettabytes globally (that's more than all the grains of sand on beaches and the desert combined)—highlights a critical vulnerability in how organizations handle their information assets. Compounding this issue is what a recent Harvard Business Review analysis revealed[4]: only about 27% of

[3] DATAVERSITY. "Smart Data Fingerprinting: The Answer to Data Management Challenges." April 30, 2024. https://www.dataversity.net/.

[4] Redman, Thomas C. "Bad Data Costs the U.S. $3 Trillion a Year." *Harvard Business Review*, September 22, 2016. https://hbr.org/.

companies consistently maintain high data quality across their operations, leading to annual losses exceeding $3.1 trillion due to bad data practices. In an era where both Applied and Generative AI rely heavily on vast datasets for training and inference, these deficiencies aren't just inefficiencies; they're ticking time bombs for security breaches, compliance failures, eroded trust, and poor-quality outputs from any model (sorry for the bad news).

This leads us to the next section. As it stands, there's no shortage of friction points that security and governance specialists face today when working with enterprise data. These challenges are amplified by the rapid integration of AI technologies, which require real-time access to diverse data sources while introducing new risks, such as model poisoning and prompt injections. Diving deeper, here are some common issues CDOs, CISOs, CIOs, and CAIOs face today:

1. **Shadow data and poor visibility as to where sensitive or valuable data lives.** Shadow data (information stored outside of sanctioned data stores) continues to proliferate, often in cloud services, personal devices, or unauthorized, unsanctioned AI tools. A 2025 report from Reco on Shadow AI[5] reveals that 27% of small business employees use risky AI applications, increasing data

---

[5] Reco. *The State of Shadow AI Report.* August 5, 2025. https://www.reco.ai/.

exposure and potentially adding $670,000 in breach costs per incident. Enterprises struggle with visibility because traditional discovery tools often fail to scan fragmented environments, resulting in blind spots where sensitive, personally identifiable information (PII) or intellectual property may lurk undetected. For instance, in hybrid cloud setups, data might migrate between on-premises servers and SaaS platforms without logging, making it nearly impossible for CISOs to map assets comprehensively or CDOs to govern properly. This lack of visibility not only heightens the risk of data leaks but also complicates incident response, as teams can't quickly isolate compromised elements.

2. **Data quality issues that degrade model performance and outputs.** Poor data quality manifests in inaccuracies, duplicates, and biases that infiltrate AI models, leading to flawed predictions or discriminatory outcomes. Harvard Business Review's 2025 insights[6] emphasize that unstructured data, which constitutes 80% of enterprise information, requires rigorous quality improvements to unlock AI value—yet many organizations overlook this, leading to "garbage in,

---

[6] Redman, Thomas C. "To Create Value with AI, Improve the Quality of Your Unstructured Data." *Harvard Business Review*, May 20, 2025. https://hbr.org/2025/05/to-create-value-with-ai-improve-the-quality-of-your-unstructured-data

garbage out" scenarios. For example, in financial services, incomplete transaction datasets can train fraud detection models that miss subtle anomalies, exposing firms to losses. Without automated quality checks, such as using machine learning for anomaly detection, models can drift over time, resulting in reduced accuracy and trustworthiness.

3. **Fragmented governance policies across platforms and business units.** In the AI era, data governance often resembles a patchwork quilt, with inconsistent policies between departments, clouds, and regions. Accenture's State of Cybersecurity Resilience 2025 report[7] notes that fragmented platforms result in duplicated data, inconsistent classifications, and siloed risk assessments, which hinder AI-driven initiatives. For multinational enterprises, this fragmentation clashes with varying regulations, such as GDPR in Europe or CCPA in the U.S., creating compliance nightmares. Business units might adopt their own AI tools without centralized oversight, resulting in policy gaps that adversaries exploit. A 2025 Collibra survey[8] highlights that

---

[7] Accenture. *State of Cybersecurity Resilience 2025*. Accenture, 2025. https://www.accenture.com/us-en/insights/security/state-cybersecurity-2025

[8] Collibra. "New Survey from Collibra by The Harris Poll Reveals Top Concerns for Tech Decision-Makers." April 30, 2025. https://www.collibra.com/.

siloed data assets prevent discovery and trust, with 74% of leaders citing transparent governance as a top need for AI success.

4. **Difficulty classifying unstructured data: a prime data input for GenAI.** Unstructured data— emails, videos, documents, and social media —grows at a 61% compound annual rate (per IDC[9]), but classifying it remains arduous due to its lack of predefined structure. Yet, many GenAI models thrive on this data for natural language processing or image recognition. But without advanced AI-native classification, sensitive elements such as trade secrets slip through. A 2025 DATAVERSITY report[10] warns that unstructured data hinders the safe deployment of GenAI, recommending sanitization and redaction to mitigate risks. Failure here can lead to biases in AI outputs or inadvertent exposure of confidential information during model training.

5. **Inadequate lineage and provenance tracking: complicating audit and compliance.** Data lineage, the traceability of data from its origin to

[9] Reinsel, David, John Gantz, and John Rydning. *Data Age 2025: The Evolution of Data to Life-Critical*. Framingham, MA: International Data Corporation (IDC), sponsored by Seagate, March 2017. https://www.seagate.com/www-content/our-story/trends/files/Seagate-WP-DataAge2025-March-2017.pdf.

[10] DATAVERSITY. "Unstructured Data Hinders Safe GenAI Deployment." September 23, 2024. https://www.dataversity.net/.

consumption, is essential for AI accountability; yet, many systems lack robust tracking. Oval Edge's 2025 guide to data lineage tools[11] emphasizes that without it, organizations can't perform forensic audits or prove compliance under frameworks such as IST AI RMF. In AI workflows, where data transforms through multiple layers, poor provenance can obscure biases or manipulations, as seen in model drift cases. Reports from Atlan[12] indicate that regulatory data lineage tracking is crucial for 2025 audits, as it helps map journeys and ensure data integrity. Without automated tools, which track lineage during model training, enterprises face hefty fines and reputational damage.

All these issues culminate in data drag, lag, and quality degradation, slowing AI adoption and inflating costs. But enough bad news! While these issues are a headache for CDOs, CISOs, CIOs, and CAIOs alike, they are also an opportunity to stand out when solved. You can do this by being maniacally focused on addressing these blind spots and leveraging AI-native capabilities to combat AI demands. Tech executives can turn data into a strategic asset, fostering innovation while bolstering defenses.

---

11 Oval Edge. "Top 25 Data Lineage Tools in 2025: Features, Strengths & Comparisons." August 6, 2025. https://www.ovaledge.com/.
12 Atlan. "Regulatory Data Lineage Tracking for Audit Success in 2025." May 2, 2025. https://atlan.com/.

However, to seize this opportunity, we must first eliminate blind spots, enforce intelligent controls, and reset our mindsets from reactive (responding after threats emerge) to responsive (anticipating and adapting in real-time). But how? We first accomplish this through a series of **Mental Model** shifts that need to take place. They include:

1. **Pivot from Content Scanning to Intent Detection:** Instead of solely focusing on the data being generated or transmitted, we need to develop capabilities to analyze the underlying *intent* behind the transmitted data, identifying potential misuse or unauthorized access to sensitive information. Intent detection uses natural language understanding (NLU) to discern malicious goals, such as extracting PII through clever queries. This proactive approach reduces risks in GenAI interactions, ensuring prompts align with ethical and security standards, and is vital as OWASP reports a rise in prompt exploits. Many of the hyperscalers have guardrails built in to ensure prompts are 'secure', but that's not what we're discussing here. What we're talking about is near-real-time prompt provenance and prompt security, capabilities like Cyera's AI Shield feature that assist in intent detection as it's happening. This is not to be confused with guardrails; instead, this is an operational process that allows you to execute on near- real-time intent detection.

2. **From Endpoint Guardrails to Model Interaction Monitoring:** Traditional endpoint security measures, like device controls and network restrictions, are insufficient against AI-specific threats. Organizations must implement comprehensive monitoring that tracks every interaction between employees and AI models, ensuring alignment with policies. Real-time observability to detect anomalies, such as data exfiltration, during inference has now become an operational imperative. So, explore the variety of tools that offer AI observability, so you can log queries, responses, and metadata —enabling rapid threat response and compliance.

3. **From Access Control to Session Auditing:** AI workflows require granular access management that goes beyond static controls. Deploying AI to aid audit workflows by flagging deviations in real-time is now the shift that needs to be considered to keep pace with the business and its requests. This shift helps detect insider threats or anomalies, such as unusual prompt patterns, and supports forensic reviews under regulations, while maintaining speed.

Needless to say, the CXO playbook has changed, and Moore's Law is in full effect— so how do we facilitate this pivot within our organizations and teams? Moore's Law, which explains exponential growth in computing power,

now applies to AI advancements, accelerating threats and opportunities alike, and the AI flywheel spins faster and faster, demanding CXOs to integrate security as strategy, not as an afterthought.

To begin pivoting into a responsive mindset, we've established this 7-step journey map, designed as a practical roadmap for CXOs and their teams. Each step builds on the last, fostering iterative progress while allowing adaptation to your organization's culture, industry, sector, and all its nuances— whether in finance's regulatory-heavy environment or tech's agile innovation.

1. **Assess Current State**: This step is not rocket science; we all know to begin here. However, conducting a comprehensive audit of your data landscape and using tools like Cyera to map shadow data, quality gaps, and governance fragmentation in a fast and accurate manner is now the operational imperative. We can't approach these assessments and audits as we have in the past. They can't be time-consuming or resource-heavy. Lean into the advancements of these tools to speed this process up and answer these critical questions in days, not years:

   - How much data do we have?
   - How much of that is dark data?
   - How much of that is sensitive data?
   - How much of that is confidential data?

- How much of that is currently violating GRC policies?
- What's our inventory of sanctioned vs. unsanctioned AI Tools?

What was once impossible to answer is now within reach. Lean in and find out how!

2. **Build Cross-Functional Alliances**: Form an AI Security Council inclusive of CDOs, CISOs, CIOs, CAIOs, legal, and business leads. This ensures buy-in and aligns with responsive goals. While we have stakeholder alignment sessions and committees, this council operates a bit differently in that it's not decision-making through consensus but through consideration. The council allows for disagreements, conflicts, and differing viewpoints. As a matter of fact, it's welcomed—but the collective goal is not to block, but to enable. So, conversations are focused not on the 'NO' but on the 'HOW.'

3. **Define Responsive Principles**: Shift mindsets by documenting principles like "security by design" and "intent over content." Build and publish these tenets and train your teams on them. When disagreements or debates come up, everyone can refer back to the shared principles—reducing friction and avoiding delays.

4. **Enhance Visibility Tools**: Invest in AI-powered discovery and observability platforms, such as Cyera. It has a remarkable ability to discover and observe unstructured data and semi-structured data, the largest data sets powering your foundational models—and the most critical. This eliminates critical blind spots. Also, set KPIs for visibility (such as the ones mentioned in Chapter 2). For example, target a 90% coverage of data assets within six months.

5. **Implement Intent Detection**: Deploy prompt shields and NLP-based tools like AI Guardian to analyze user intents in real-time, reducing prompt injection risks by 70%, and automate alerts for deviations.

6. **Monitor Model Interactions**: Roll out AI observability to track all AI engagements. Establish dashboards for anomaly detection, ensuring interactions comply with policies.

7. **Test, Iterate, Measure, and Scale**: Run simulations of AI threats to refine responsive strategies. Simulate a 'user' who unintentionally (and innocently) injects confidential information into a prompt (i.e., contract number, case number) and see if your governance detects the violation. Gather feedback quarterly and track progress with metrics like the ones shared.

This 7-step journey map helps you transform challenges into strengths, positioning your organization as AI-resilient. And as I always say, think BIG, start SMALL, scale up QUICKLY and interate BOLDLY.

CHAPTER 5

Enabling Secure AI Through Data Intelligence

> According to IDC's 2025 projections[13], global data creation will surpass 175 zettabytes, with 90% being unstructured—emails, documents, and multimedia that are notoriously hard to classify and govern. A 2025 Harvard Business Review[14] report states that only 27% of enterprises maintain consistent data quality, and this costs businesses $3.1 trillion annually due to poor data practices.

In the AI era, data is the lifeblood of innovation, but its potential hinges on a concept we call **Data Intelligence** —

---

[13] Reinsel, David, John Gantz, and John Rydning. *Data Age 2025: The Evolution of Data to Life-Critical*. Framingham, MA: International Data Corporation (IDC), sponsored by Seagate, March 2017. https://www.seagate.com/www-content/our-story/trends/files/Seagate-WP-DataAge2025-March-2017.pdf.

[14] Redman, Thomas C. "Bad Data Costs the U.S. $3 Trill *usiness Review*, September 22, 2016. https://hbr.org/.

the ability to understand, govern, and secure data at scale, as we first mentioned in Chapter 1. Without this foundation, AI deployments risk becoming liabilities, exposing organizations to compliance violations, breaches, or biased outputs that erode trust. For CDOs, CISOs, and other C-level executives and their teams, the mandate is clear: transform data from a chaotic liability into a strategic asset. This chapter explores how CDOs, CISOs, CIOs, and CAIOs can leverage a practical framework for data intelligence and integrate cutting-edge solutions to address modern enterprise challenges.

As mentioned above, IDC's 2025 projections indicate that global data creation will surpass 175 zettabytes by the end of the year, with 90% being unstructured: emails, documents, and multimedia that are notoriously difficult to classify and govern, costing businesses $3.1 trillion annually. For AI, which thrives on vast, diverse datasets, these gaps translate into flawed models, regulatory fines, or reputational damage. For instance, a financial institution using unclassified customer data in a GenAI model risks leaking PII, while a healthcare provider might produce biased patient diagnoses from low-quality datasets. Can you imagine working nearly a year to create an amazing AI product for customers to leverage, only to have one bad incident stop the entire initiative? It's heartbreaking, and, truth be told it's happened to me twice! That's why it's important to remember what you do and the day-to-day decisions you're making matter!

You are front-office architects of AI-driven value. By harnessing the concept of *Data Intelligence*, a broader, bigger ecosystem play, you can now align innovation with protection, turning AI into a competitive differentiator through these five critical areas:

1. **Continuous Data Discovery and Classification Across Environments**: This is likely the most crucial step in all the frameworks and models presented. Your enterprises operate across hybrid ecosystems—cloud, on-premises, and SaaS platforms like AWS, Salesforce, or Snowflake. Data is EVERYWHERE! So, you have to use AI-native discovery and classification tools to combat the agitation AI has created on governance and security. Using AI-powered engines, like Cyera, scan these environments in real-time, identifying sensitive, confidential, or toxic combination data (e.g., PII, intellectual property) with 98% accuracy (per 2025 benchmarks[15]). Unlike traditional scanners, platforms like Cyera utilize AI-native capabilities, which enable them to discover and classify unstructured data, such as PDFs or free-text fields, ensuring that no dataset is overlooked with minimal false positives and high accuracy. I know this is coming off as a product pitch, and I'm trying my best to avoid that. So, I encourage you all

---

[15] Cyera internal data

to test and go toe-to-toe with your tools in the tech stack, and you'll see firsthand the massive difference in time, quality, cost, and accuracy. This distinction truly matters, enabling your organization to keep pace with evolving business needs and security risks.

2. **Dynamic Governance Policies Based on Sensitivity, Criticality, and Usage**: Static governance policies falter in AI's dynamic landscape. So, it's essential you put in place adaptive policies that adjust based on data attributes: sensitive (e.g., financial records vs. public data), critical (e.g., mission-critical vs. archival), and usage patterns (e.g., accessed by GenAI vs. analytics). For example, Cyera's policy engine can restrict a dataset containing health records from being used in a public-facing chatbot, reducing exposure risks by 70%, according to 2025 Cyera case studies[16]. This ensures that governance evolves in tandem with AI workloads, striking a balance between innovation and compliance. What all of this means is that what you document and outline can't be static and published into a SharePoint drive for all to follow; those days are long gone, and this is not the way to govern in the new age. You need the ability to have a system that

_____

[16] Cyera internal data

1) monitors in near real-time data activity violating your policies, and 2) automatically expands your policies with new policies based on what it is observing with user activity, sanctioned use, and unsanctioned use, and makes recommendations that you then approve. This is critical because you don't know what you don't know, and this can help with blind spots.

3. **Embedding Lineage, Context, and Usage Metadata into Workflows**: Trust and compliance depend on knowing where data comes from and how it's used. Lineage tracking maps data from its source through AI model training and inference, embedding metadata like origin, transformations, and access history. This supports forensic audits under frameworks like NIST AI RMF and enables rapid response to breaches. For instance, if a model outputs sensitive data, your runtime API should trace it back to the original dataset, identifying whether it was improperly accessed. This transparency is crucial for proving compliance and mitigating risks.

4. **Providing Safe Sandboxes for GenAI Innovation**: AI teams need environments to experiment without compromising security. Cyera's secure sandboxes allow developers to work with curated, compliant datasets, stripped of sensitive elements via automated redaction. These sandboxes integrate

with platforms like Databricks or Azure, enabling GenAI prototyping while enforcing guardrails.

In short, if data isn't classified, governed, or well-understood, AI becomes a risk, not a differentiator, and unclassified data fuels model errors, breaches, and poor understanding of how your AI capabilities are generated in the first place.

So, how does all this operationalize **Data Intelligence**? And where do you start? Well, you first assess where you are with where you ought to be, given the demands of the business. You can do this by using a structured framework to evaluate, examine, and understand where you are in the **Data Intelligence** maturity model. It consists of five maturity levels, each with specific actions and metrics, tailored for CXOs navigating the complexities of AI.

Level 1: Fragmented (Reactive Baseline)

- **Characteristics**: Data is siloed, governance is siloed and static, with minimal classification (e.g., <30% of data tagged). Shadow data proliferates, and lineage tracking is absent. AI models risk using unvetted datasets, which can lead to breaches or biases.
- **Actions**: Deploy a discovery engine to map all data assets across cloud, on-premises, and SaaS. Start with high-risk areas (e.g., customer, product). Then,

set baseline KPIs, like reducing unclassified data by 20% in six months.

- **Integration**: Use an automated AI-driven classification engine, like Cyera, to tag confidential and sensitive data, achieving 95% accuracy. Generate visibility reports to identify shadow data risks.
- **Metric**: Percentage of data assets classified (target: 40% within Q1).

Level 2: Defined (Governance Foundations)

- **Characteristics**: Data is siloed, but governance policies exist; however, they are static and inconsistent across platforms. Data quality issues persist, impacting AI model performance.
- **Actions**: Implement Cyera's dynamic policy engine to enforce rules based on data sensitivity and usage. Begin quality checks using AI-driven anomaly detection. Train teams on governance best practices.
- **Integration**: Leverage policy templates to align with GDPR, CCPA, or NIST standards. Automate data quality scans to flag duplicates or biases, reducing errors by 50%, per 2025 benchmarks[17].

---

[17] Cyera internal data

- **Metric**: Governance policy coverage (target: 60% of data under dynamic policies).

Level 3: Managed (Proactive Integration)

- **Characteristics**: Data lineage and metadata are partially embedded, enabling basic compliance. Teams experiment in secure sandboxes, but scalability is limited.
- **Actions**: Use Cyera to embed lineage tracking in AI workflows, ensuring traceability for audits. Scale sandboxes for broader AI teams, focusing on high-impact use cases like personalization or fraud detection.
- **Integration**: Deploy lineage tools to map data journeys, supporting forensic audits. Enable sandboxes with redacted datasets, reducing exposure risks by 60%.
- **Metric**: Lineage Traceability Score (target: 70% of AI outputs traceable).

Level 4: Optimized (Responsive Ecosystem)

- **Characteristics**: Data intelligence is embedded enterprise-wide, with real-time monitoring and adaptive governance. Collaboration between CDOs, CISOs, and AI teams drives innovation and security, and data governance talks are now limited as a

result of the trust in what's in place—leading to faster time to market for AI products and capabilities (the fun stuff).

- **Actions**: Expand runtime APIs for real-time prompt monitoring and intent detection. Integrate session auditing to track user interactions with AI models. Optimize data pipelines for performance and compliance.
- **Integration**: Use the runtime alerts to detect prompt leakage (e.g., 95% detection rate) and session auditing for zero-trust compliance. Optimize datasets for AI training, improving model accuracy by 30%.
- **Metric**: Prompt Leakage Rate (target: <5% sensitive data in prompts).

Level 5: Transformative (AI-Driven Leadership)

- **Characteristics**: Data intelligence powers AI as a strategic asset, providing complete visibility, governance, and trust. The organization leads in responsible AI, with measurable ROI and compliance.
- **Actions**: Institutionalize dashboards for cross-functional alignment, tying AI outcomes to revenue or cost savings. Share success stories to build a culture of responsible AI.

- **Integration**: Leverage analytics to quantify AI ROI (e.g., 15% revenue lift from personalization models). Use dashboards to report to boards, reinforcing trust.
- **Metric**: AI Attributable Revenue Growth (target: 10–20% tied to AI initiatives).

This maturity model enables CXOs to assess their current state, set clear goals, and track progress with measurable outcomes. While CXOs can use a variety of tools in the tech stack to accomplish some of these activities, organizations can accelerate their journey by embedding *Data Intelligence* into every AI workflow, in real time and dynamically, when using best-of-breed platforms.

CHAPTER 6

Picking the Right Partners and Tools

> *Only 11% of companies have successfully scaled AI beyond pilots; the rest cite integration with existing processes as the top barrier (McKinsey & Company, 2022[18]).*

You've built your Data Intelligence foundation and mapped your security framework, and now you think you can relax. The reality is, the real work is about to begin. Why? Because if you choose the wrong partner in this journey with you, AI can pivot from becoming your competitive advantage to your compliance nightmare.

---

[18] Bughin, Jacques, Michael Chui, Roger Roberts, and Kate Smaje. *Moving Past Gen AI's Honeymoon Phase: Seven Hard Truths for CIOs to Get from Pilot to Scale.* McKinsey & Company, July 11, 2024. https://www.mckinsey.com/capabilities/mckinsey-digital/our-insights/moving-past-gen-ais-honeymoon-phase-seven-hard-truths-for-cios-to-get-from-pilot-to-scale.

When it comes to picking partners and tools, the marketplace is crowded, fast-moving, and, at times, dangerously misleading. Every solution claims transformation. Every platform promises acceleration. Yet many vendors fall short, and organizations stall because solutions are misaligned with existing systems and teams. So while picking partners and tools is a very personal decision, we created a framework for you to leverage, so you have a compass when navigating the AI provider landscape.

Step 1: Clarity vs. Clouds (Cut Through the Noise)

When the time comes to interview, research, and explore providers, follow these steps:

- **Ask:** The provider to explain, in plain terms, how the system works and the limitations across circumstances.
- **Listen For:** Buzzwords versus transparent explanations of data sources, model training processes, and limitations. Do they make consistent references to "proprietary algorithms" or "intelligent black boxes", or do they avoid the buzzwords? How far do they go into explaining on-demand availability, documentation on processes, and use cases for compliance?
- **Signal:** Most employees of a vendor don't know how the technology works, so it's your job to parse out

fact from fiction and identify the clearest way to tell a partner from an opportunist. What cannot be explained clearly cannot be considered credible.

Step 2: Impact Over Features (Reality Check)

When you've narrowed your options and you now have a select group of partners and technologies you're exploring, you dig a bit deeper and go into the following details:

- **Ask:** What measurable outcomes has this solution delivered for similar organizations?
- **Listen For:** Concrete results such as reduced incident response time, increased model accuracy, stronger data classification, or tangible cost savings. Don't fall for feature lists or generic promises of "transformation"; listen for proof points with credible sources.
- **Signal:** Features without evidence of impact often end up as shelfware, so you must focus on the outcomes they share over the menu of feature options.

Step 3: Alignment Check (The Litmus Test)

From here, you're ensuring solutions integrate and align with what your company and culture have designated as their governance processes and protocols. So now you:

- **Ask:** How does the offering fit with your governance, compliance, and workforce realities?
- **Listen For:** Specific references to relevant regulations, industry standards, and operational workflows. Avoid one-size-fits-all answers that ignore enterprise context. Does the solution help future-proof against emerging AI regulations?
- **Signal:** Since misalignment creates new risk, this is where the rubber meets the road, and you're likely going to filter out most partners and tech stacks.

Step 4: Proof of Integrity (It's in the Pudding)

So now you're down to 2 or 3 options, and this is where specificity and proof points become your best friend. In this step, you will:

- **Ask:** To see evidence that validates the vendor's claims; explore case studies, benchmarks, and references from credible sources.
- **Listen For:** Willingness to connect executives with peer organizations, verifiable data points, and references of success stories; get past the selective testimonials and glossy marketing collateral as quickly as you can.
- **Signal:** Pay attention to those who demonstrate openness by telling you up front the lessons learned, versus those who deflect and resist

scrutiny. Those who resist, avoid. Those with transparency are your better partners; they will share the burden with you.

Step 5: Longevity (In it for the Long Haul)

This is the hardest step, because it's in this step that you may have to face the tough decision of selecting a startup that's AI-Native with an amazing capability, but has raised only a few million, or go with the known entity, a juggernaut platform, with newly embedded AI that meets part of your needs. What you can do is:

- **Ask:** What does success look like six to twelve months after go-live?
- **Listen For:** Plans for retraining, ongoing support, alignment with product roadmaps, and how they plan to pivot if assumptions don't hold true or the market changes.
- **Signal:** Since the AI vendor landscape changes every 6 months, see how they share their plans. If one vendor illustrates a long-term strategy, understand that there are a lot of assumptions baked into it and listen for pivot plans. If a vendor plans in quarters, it could mean they operate in a truly agile way, where they measure velocity of deployments by weekly increments, so now you need to hear about their vision for the year, and if

it's based on customer and market input to ensure product-market fit.

The reality is, choosing partners is tricky no matter what. But if you follow these five steps in our framework, you'll be better positioned to make a decision that sets you up for a higher probability of success during your deployment. The traits you want to look for include:

- Upfront visibility, like sharing an incident of a failed deployment and the reasons why.
- Comfort in credibility, because they have the proof points to back up their claims.
- Clarity and conciseness—you don't have time for fluff and buzzwords.

Also, remember, it's not one-size-fits-all in the AI game. You can choose to play it safe by going with the known entity that just recently launched their AI feature, but you may never get ahead. Therefore, you'll always be reacting to business needs because this entity moves slowly. However, everyone finds comfort with the bigger players. Nor can you choose the shiny startup whose capabilities are amazing but may not be around in one year. But, on the flip side, they are so good at what they do that, even if it's only for one year, they put you ahead of the game.

The game of picking your partners is both art and science; there's no way around it.

CHAPTER 7

The Evolving Role of the C-Suite

> *2025 global CIO survey by Gartner[19] found that over 80% of CIOs plan to invest in cybersecurity and risk management, generative AI and ML, data analytics, and integration & APIs—reflecting that executive boards increasingly see AI and cybersecurity as business-critical priorities, not just technical issues.*

Throughout this eBook, we have discussed several frameworks and models for you to leverage; however, it wouldn't be prudent of us to ignore the obvious: the evolution of our roles as C-suite executives. Whether it be CDOs, CISOs, CIOs, or the emerging Chief AI Officers (CAIOs), our roles and our teams are now strategic architects, navigating a landscape where AI amplifies both

---

[19] https://www.gartner.com/en/newsroom/press-releases/2024-10-22-gartner-survey-reveals-that-only-48-percent-of-digital-initiatives-meet-or-exceed-their-business-outcome-targets

opportunities and risks. As organizations integrate AI into core processes, we are being asked to solve and evolve the conundrum of balancing innovation and value creation, with security and compliance. This chapter examines how these roles are being reshaped and offers a roadmap for thriving in the AI-driven enterprise.

AI has perhaps made our roles more complicated; we're being asked to do things that have never been done before and at a pace that frankly isn't realistic. This now elevates CDOs, CISOs, CIOs, and CAIOs to more strategic roles, requiring them to align AI initiatives with revenue goals, cost efficiencies, and regulatory mandates. The stakes are high. As IBM stated in 2025[20], the cost of a data breach or security concerns is now up to $4.88 million on average, with AI-related incidents adding 20% to recovery costs due to complexities like model poisoning or prompt injections.

Also the traditional boundaries across the C-Suite and their roles are blurring. CDOs, once focused on data management and governance, now drive AI-driven business outcomes. CISOs, no longer just cybersecurity gatekeepers, oversee AI model security and compliance. CIOs shift from IT infrastructure to enabling AI-powered transformation, while CAIOs emerge to orchestrate AI strategy. This convergence demands new skills—along

---

[20] IBM Security. *Cost of a Data Breach Report 2025*. Armonk, NY: IBM Corporation, 2025. https://www.ibm.com/reports/data-breach

with unprecedented collaboration. Let's discuss each role in detail:

Chief Data Officer (CDO): *from Data Steward to AI Value Creator.* CDOs are transitioning from data governance to unlocking AI's business potential. Their role now includes:

- **Orchestrating AI Data Pipelines**: Ensuring high-quality, compliant datasets for AI models, leveraging data classification and lineage tracking as their pillars to success, and developing near-real-time governance protocols.
- **Driving ROI**: Linking datasets to outcomes, like a 15% sales uplift from personalization models.
- **Ethical Stewardship**: Mitigating biases, ensuring accuracy by providing good quality and diverse data. CDOs must master AI-specific tools (e.g., Snowflake for data prep) and partner with CAIOs to prioritize high-value use cases, such as predictive analytics in manufacturing or fraud detection in finance.

Chief Information Security Officer (CISO): *from Gatekeeper to AI Risk Strategist.* CISOs are evolving into strategic risk managers, with responsibilities expanding to:

- **Securing AI Models**: Protecting against threats like prompt injections and runtime API abuse.

- **Regulatory Compliance**: Ensuring AI aligns with GDPR, CCPA, and NIST AI RMF, which is why 82% of CISOs now report to CEOs[21].
- **Real-Time Monitoring**: Technical expertise (e.g., CISSP certifications) with boardroom communication, translating AI risks into business impacts.

Chief Information Officer (CIO): *from IT Leader to AI Innovation Enabler*. CIOs are shifting from infrastructure oversight to enabling AI-driven transformation:

- **Strategic Alignment**: Integrating AI into IT strategies, like cloud migrations supporting GenAI, with 90% of enterprises adopting hybrid clouds by 2027(per Gartner 2025[22]).
- **Vendor Management**: Negotiating AI tool contracts to optimize costs

---

[21] Splunk Inc., in collaboration with Oxford Economics. *The CISO Report 2025: CISOs Gain Influence in the C-Suite and Boardrooms Worldwide*. San Francisco: Splunk Inc., January 23, 2025. https://www.splunk.com/en_us/newsroom/press-releases/2025/splunk-report-cisos-gain-influence-in-the-c-suite-and-boardrooms-worldwide.html

[22] Gartner, Inc. "Gartner Forecasts Worldwide Public Cloud End-User Spending to Total $723 Billion in 2025." Press release, November 19, 2024. https://www.gartner.com/en/newsroom/press-releases/2024-11-19-gartner-forecasts-worldwide-public-cloud-end-user-spending-to-total-723-billion-dollars-in-2025

- **Cross-Functional Leadership**: Collaborating with CDOs and CISOs to embed AI securely. CIOs need business acumen and AI literacy to drive initiatives like automation.

Chief AI Officer (CAIO): *the New Architect of AI Strategy*. The CAIO, an emerging role, orchestrates AI adoption:

- **AI Roadmap Development**: Defining use cases, like personalization or supply chain optimization, and setting the stage for how companies embed AI into their operational workflows.
- **Ethical AI Governance**: Ensuring fairness and transparency, and demand ethical AI (per Edelman's 2025 Trust Barometer[23]).
- **Team Alignment**: Bridging CDOs, CISOs, and CIOs to prioritize AI projects. CAIOs require a blend of technical, strategic, and ethical skills, often holding advanced degrees in AI or data science.

To navigate this new landscape and thrive in an AI-driven enterprise, CDOs, CISOs, CIOs, and CAIOs need to come together and align on the division of labor among them, co-owning a unified AI vision that sets shared goals while preparing their teams and the workforce collectively. Easier said than done, right? Nevertheless, it has to be pursued. Why? Because in the AI-driven enterprise, the

---

[23] Edelman. *2025 Edelman Trust Barometer*. Edelman, January 2025. https://www.edelman.com/trust/2025/trust-barometer.

most sophisticated metrics, frameworks, and governance strategies are only as effective as the teams executing them. As such, CXO alignment and workforce preparation are the linchpin that turns theoretical AI data security into practical resilience.

In this next section, we will delve into how CXOs can prepare their teams for the complexities of AI, defining workforce preparation, outlining actionable steps with clear objectives, timelines, and resources, and examining the risks, considerations, and transformative benefits of implementing this process.

CHAPTER 8

Workforce Preparation in the AI Governance Landscape

Workforce preparation refers to the systematic process of assessing, upskilling, reskilling, and culturally aligning teams to handle the intersection of AI technologies, data management, and security threats. In the context of AI governance, it's about building a workforce that not only understands traditional cybersecurity but also grasps AI-specific risks, such as prompt injections, model poisoning, data biases, and regulatory compliance under frameworks like NIST AI RMF or GDPR. This involves fostering a "security-aware culture" where employees—from data engineers to business analysts—proactively mitigate risks while leveraging AI for innovation. So, it's key to know that preparation extends beyond technical training—it encompasses ethical awareness, cross-functional collaboration, and adaptive mindsets, ensuring teams can effectively respond to evolving threats.

CYERA

So where should CXOs start? A good approach is to view workforce preparation as a phased initiative. The seven steps below are drawn from practical experience and resources such as *Your AI Survival Guide: Bruised Knees, Scraped Elbows, and Lessons Learned from Real-World Deployments*[24].

Step 1: Conduct a Skills Gap Assessment

- **Objective**: Identify current competencies in AI governance and ethics to pinpoint deficiencies, such as a lack of knowledge in AI model auditing or data lineage tracking. This baseline ensures targeted preparation, aligning with organizational AI goals.
- **Timeline**: Give these 1-2 months, starting immediately.
- **Resources Needed**: Use internal surveys, tools, consultants, or platforms to assess and lean into HR for anonymous feedback to encourage participation.

Step 2: Develop a Customized Training Curriculum

- **Objective**: Create tailored programs covering AI fundamentals, security best practices (e.g., intent detection, session auditing), and governance

---

[24] Rashidi, Sol. *Your AI Survival Guide: Bruised Knees, Scraped Elbows, and Lessons Learned from Real-World Deployments*. Hoboken, NJ: John Wiley & Sons, 2024.

frameworks (e.g., NIST, which is excellent, but discern if it's overkill for your organization). The goal is to bridge gaps, empowering teams to implement metrics such as Prompt Leakage Rate or Lineage Traceability Score, as introduced in earlier chapters.

- **Timeline**: Allocate 2-4 months for this phase, overlapping with the assessment. This is not a linear step; it can be done in parallel with step 1.
- **Resources Needed**: Online platforms (like Coursera or Cyera's AI Security School) for AI governance training. You can also lean into specialized modules on data classification and prompt poisoning. Try to allocate 20-30 hours per employee for initial training.

Step 3: Implement Hands-On Workshops and Simulations

- **Objective**: Provide practical experience through scenarios like simulating AI breaches or using sandboxes for GenAI experimentation, fostering application of knowledge to real-world threats like data poisoning.
- **Timeline**: Allocate 2-3 months, post-curriculum development.

- **Resources Needed**: Use simulation software, secure sandboxes, facilitators from organizations like SANS Institute, and internal hackathons to facilitate the hands-on experience.

Step 4: Foster Cross-Functional Collaboration and Mentoring

- **Objective**: Break silos by pairing data, security, and AI teams, promoting knowledge sharing and alignment on responsibilities.
- **Timeline**: This is an ongoing exercise that becomes an organizational operating model.

Step 5: Integrate Continuous Learning and Certification

- **Objective**: Embed lifelong learning to keep pace with AI evolutions, such as new threats from agentic AI, certifications like AI Security School from Cyera, and use case updates from other industries as the dos and don'ts for your training.
- **Timeline**: 3-4 months initial rollout, then perpetual.

Step 6: Establish Performance Metrics and Feedback Loops

- **Objective**: Measure preparation effectiveness through KPIs like training completion rates or

reduced incident response times, iterating based on feedback to refine programs.
- **Timeline**: 6-9 months, with quarterly reviews.
- **Resources Needed**: Analytics tools integrated with HR systems.

Step 7: Cultivate a Responsive Security Culture

- **Objective**: Shift mindsets from reactive to responsive, embedding AI ethics and security as core values.
- **Timeline**: Evolving indefinitely.
- **Resources Needed**: Leadership workshops, communication campaigns, and reinforcement of governance in daily workflows.

While workforce preparation will be fraught with challenges because of the resistance to change, and employees' sentiment towards AI training as either overwhelming or irrelevant, CXOs can't give up if they first see low engagement or low knowledge retention. Because if we don't prepare the workforce, the "skills chasm" exacerbates breaches and security exposure.

Comprehensive workforce preparation programs can exceed $100,000 annually, straining budgets in resource-constrained environments. But, if you include it as an embedded cost in doing AI, and not as a separate line item (don't make this piece negotiable), you'll have the budget to train appropriately.

Strategically, a responsive workforce becomes a competitive differentiator, not just the latest use case on a fancy foundational model. In the long term, this preparation transforms AI from a challenge into a sustainable asset. In the grand architecture of AI governance, workforce preparation stands as the most crucial step; that's why we saved the best for last. Without teams equipped to act, the metrics, strategies, and governance outlined in the prior chapters remain blueprints gathering dust, unable to withstand the realities of implementation and evolving threats.

CHAPTER 9

Conclusion: Build the Grid, Not Just the Gadgets

AI is no longer a side project or a shiny pilot; it's becoming the operating system of the modern enterprise. Throughout this eBook, we've stressed a simple, non-negotiable truth: **AI value only scales when visibility, classification, and control are engineered into the fabric of how data moves and how models behave.** That's why we anchored on three converging pillars:

- **Security as a system** (the 7-Phase AI Security Framework)
- **Value as a scoreboard** (the AI-Security ROI Metric System)
- **People are the true platform** (workforce preparation as the decisive advantage)

Together, they form a closed-loop governance model that turns intent into impact. Data is the power grid, AI is the appliance layer, and your governance is the grid operator

—balancing load, preventing outages, and expanding capacity with confidence.

When you head back to work after this eBook, what's different? Here's what will change starting Monday!

We govern flows, not just files. Shadow data, prompt streams, and API calls are now first-class citizens in our security model. Lineage and session auditing become as core as encryption and MFA.

1. **We measure trust like we measure revenue.** Time-to-Trust (T2T), Prompt Leakage Rate, Lineage Traceability, Model Access Governance Scores—these are not academic. They are board-ready indicators of whether AI is safe to scale.
2. **We shift from reactive to responsive.** From content scanning to intent detection, from endpoint guardrails to model-interaction monitoring, from static access control to continuous session auditing.
3. **We professionalize AI operations.** The 7-Phase framework moves us from scattered controls to an engineered control plane that spans sourcing → infrastructure → in-transit protections → API security → model provenance → incident response → continuous monitoring.
4. **We invest in the workforce as the multiplier.** Upskilling beats reorgs. A prepared team reduces breach probability, shortens response time, and unlocks better models with cleaner data.

Executive Takeaways (Print This!)

- **Adopt the 7-Phase AI Security Framework** as your backbone. Treat gaps as risks on a register with owners and dates, not as "future work."
- **Run on metrics.** Stand up a live dashboard for: Data-to-Model Yield, Prompt Leakage Rate, Lineage Traceability, Model Access Governance, T2T, Model Deployment Velocity, Retraining Frequency, AI-Attributable Revenue, and Cost-to-Serve Reduction.
- **Instrument runtime.** Log and analyze every AI session (user, prompt, response, tool call, RAG source). If you can't see it, you can't secure it—or improve it.
- **Automate classification.** Manual tagging will never keep pace. Auto-classify with high-precision models; route handling and retention by sensitivity.
- **Prove provenance.** Embed lineage and chain-of-custody from data acquisition through inference. If an output surprises you, you must be able to trace it back in seconds—not days.
- **Reduce blast radius.** Segment networks, isolate model services, constrain RAG corpora, and enforce least privilege on API keys and agents.
- **Practice incident response for AI.** Red-team prompt injection, jailbreaks, data exfil via outputs, and model drift. Tabletop exercises should include Legal, Comms, and Risk.

- **Codify "intent over content."** Move beyond keyword DLP to intent detection on prompts and responses to catch indirect leakage and policy evasion.
- **Close the talent loop.** Fund a standing curriculum on AI security, data governance, and ethics; certify roles; reward completion; repeat quarterly.
- **Be boring about compliance.** Map controls to GDPR/CCPA/NIST/sector regs; automate evidence collection; produce audit-ready reports monthly.

A 90-day sprint you can start now!

- **Days 1–30: See everything.**
  Inventory AI usage (internal + 3rd-party). Turn on session auditing for prompts/responses. Launch auto-classification across cloud/SaaS. Baseline Prompt Leakage Rate, Data-to-Model Yield, and Lineage Traceability.

- **Days 31–60: Control everything.**
  Deploy guardrails at ingestion, API, and output layers. Enforce RBAC + rate limits on model endpoints. Stand up secure sandboxes with redaction for GenAI experimentation. Publish the AI Security Policy v1 (intent detection, session logging, RAG hygiene).

- **Days 61–90: Prove everything.**
  Light up the AI-Security Metric dashboard for executives. Run an AI incident simulation and capture Mean Time to Detect/Respond. Present board-level results: reduced leakage, improved T2T, and at least one production use case with tracked ROI.

The leadership message

In conclusion, we at Cyera want you to use our platform to help you every step of the way, because we know it can; we're biased for obvious reasons. But platform pitch aside, and if you're like me (always suspicious of vendors who write an eBook), don't let it take away from the frameworks, mental models, tenets, and best practices outlined in this eBook that have been provided from experience. We want you to be successful, no matter what! That's our first and foremost priority. Because we now live in a world where future leaders won't be judged by how many models they built, but how safely and meaningfully those models changed outcomes.

The **CTO/CDO/CAIO/CISO/CIO coalition** now sits at the intersection of growth, trust, and resilience. That is not a burden, but a tremendous opportunity for you, your community, and your company. This is a privilege and an opportunity.

- **If you design for visibility, classification, and control**, you earn the right to scale.
- **If you measure value and risk together**, you earn the right to invest.
- **If you train the workforce**, you earn the right to transform.

The world is watching. Your leaders are watching. Your investors are watching. Also, regulators are catching up, competitors are iterating, and customers are learning faster than ever. In this environment, **what you do matters**— not just for quarterly reviews— but for the integrity of your work and the safety of the systems you're building. So, remember, you are the stewards of the power grid for the 21st century. Adopt the frameworks. Turn on the metrics. Prepare your people. Then scale— with courage and control.

And if you'd like, ask about Cyera. We'd love to help you on this journey. We have the technical capabilities and the advisory experience because we were born 6 years ago as an AI-native company (before it became cool to say it). We were created to solve this exact problem, and we've been maniacally focused on AI Governance and Data Security in the new age since the start. We're here to help!

Final call-to-action (commitment checklist):

- Executive AI Security Council formed (meets biweekly; decisions documented).
- 7-Phase controls mapped to owners, dates, and budgets.
- Runtime session auditing is live across all sanctioned AI tools.
- Auto-classification ≥90% of high-value data; lineage enabled for AI pipelines.
- AI-Security Metric dashboard published to the C-suite.
- Quarterly AI incident simulations scheduled and resourced.
- Workforce curriculum funded and measured (completion, retention, application).

Most importantly, lead like it matters, because it does!

www.ingramcontent.com/pod-product-compliance
Lightning Source LLC
Chambersburg PA
CBHW051331220526
45468CB00004B/1584